This book is written with Love of my Father and Mother, also for my Brothers and my dear Son.

- Xiaohong

Preface

This book is presented to anyone who has concerns about cancers for one's self or for others. The intention is to provide the insight for those who are interested in understanding cancer from scientific or health point of view, and who have understanding about basic concepts of biology. The reader who didn't have some basic education on biology, it would be possible for a person to read through this book with a small degree of difficulties. Any benefit this book can bring to readers will be very rewarding to the author. I want to address that Dr. Michael Lutz has provided a great amount of pleasure of discussions and exchange in ideas on the subjects in this book.

Most part of this book was based on personal imagination or opinions, as well as public scientific literatures. It is not indented to provide suggestions for health care. In order to reduce risks and maximum the benefit, I would like to suggest the readers to use this book as a reference only.

Xiaohong Gu

Introduction

Once upon a time, I was studying Electric Engineering. I was fascinated by and absorbed in studying the most advanced mathematics, physics, system and information theory and technologies.

One day, I read a book about Norbert Wiener, the father of Cybernetics. Wiener studied both biological systems and manmade mechanic systems. He established the foundation of cybernetics theory and explained how the systems are controlled. I was fascinated by his approach and achievement.

"He must have learnt a lot from biology systems." I thought.

At that time, I knew very little about biology systems. Only thing I know at that time was how mechanic and electronic systems could run. I read a few books about Simulation of Biological Systems. It looks to me that biological systems had some intriguing designs I had never heard of. These systems imitated biological mechanisms worked better than what human designed. The ideas were so intriguing and sophisticated.

"Some super power designed these systems." I guessed.

I then felt strong urge to study biology. I wanted to work on something more superior than what a human mind can ever come up with. Another reason drove me on this new direction was that I believed that I could be happier if I studied something directly associated with people and me. I felt that I had somewhat lost in touch with real life because my mind was filling up with manmade systems all the time.

"I need to study the systems designed by God." My mind became very clear.

I mentioned my thought to one of my fellow graduate student, Li.

"Do you know anything about biology?" Li asked me

"Not much. But I can learn." I answered.

He looked at me with quite bit of doubt and a bit of concern in his eyes.

"It is a nice dream but it will be difficult to come true." Li said.

"I will give a try." I answered.

After my graduation, I received my Master of Science degree in Electronic Sciences. I then jointed the Academia Sinica. I started my journey in biology study. After I read many biology books and did quite a few experiments on marine animals, I went to Japan to study genetics and development of mouse, monkey. Everything I learnt was so fascinating to me. However, I felt that I was floating on the surface of biology. I realized that unless I understand biology at molecular level, I could only scratch the outside of any biology systems.

"I should study Biochemistry and Biophysics." I told myself.

I came to US to study with Dr. William Marzluff. In 1995, I finished my graduate study and received Doctor of Philosophy degree in Biochemistry and Biophysics from University of North Carolina in Chapel Hill. I was overjoyed when Dr. Donald McDonnell in Duke University accepted my application to work in his research lab. Donald is a leading scientist in the field of breast cancer research. I was so excited because I would work on the research projects that directly associate with me, a woman.

Before I started working with Donald, I tried to educate myself on breast cancer issues. The first book I read mentioned that 1 out of 8 woman in US have breast cancers. "Wow! That is a lot of women!" I said to myself. I was wondering why I didn't hear much about breast cancer in China.

In my real life, I enjoy chatting with women, even a lady I have never met before. However, I often feel sorry for what I heard from them. Many ladies I met or spoken to mentioned to me that either someone in their family or themselves have gotten breast cancers. As a woman scientist studying breast cancer, I constantly feel I have duty to explain to my friends and every lady in the world the nature of the disease and how can we prevent and cure the diseases.

Below is the survey published by the National Cancer Institute on breast cancer patient population study (Surveillance, Epidemiology, and End Results (SEER) Program (www.seer.cancer.gov) SEER*Stat Database: Populations - Total U.S. (1969-2004), National Cancer Institute, DCCPS, Surveillance Research Program, Cancer Statistics Branch, released April 2007.). The survey stated that: "178,480 women will be diagnosed with and 40,460 women will die of cancer of the breast in 2007"

The following chart showing that the number of incidence of breast cancer occurrences among 100,000 people during 2000-2004.

http://seer.cancer.gov/statfacts/html/breast.html

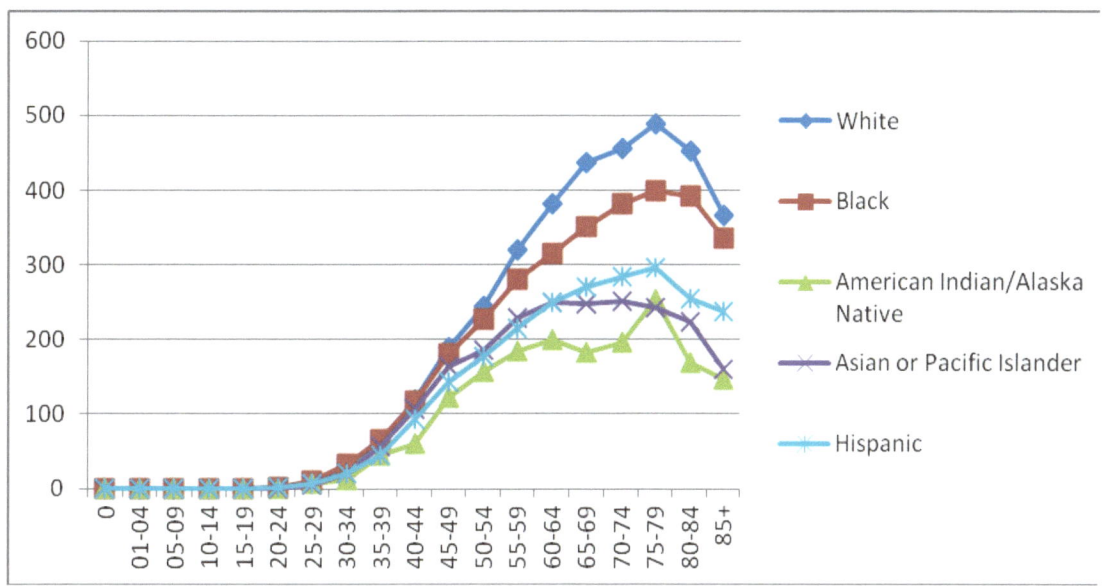

This piece of data had drawn my attention. First all, it shows that the breast cancer risk peaks at age 60 -80. This says that when we are old enough and we don't need to feed babies any more, we will get fatal disease. The second, white and black people have significantly higher risk than people of other races. No wonder so many American people I talked to either have breast cancer themselves or some of their direct family members have breast cancer. Still so many Asian, American Indian and Hispanic people have breast cancer. Apparently, no one can completely escape from breast cancer threat.

May be the survey result was a coincidence? The survey was based on millions of samples. This means that the data source size is quite large. Look at the curves: these curves are so smooth. It couldn't possibly be accidental. There must be some reasons for why the curves look like that.

I was wondered: Why do we women are punished like this?

During 4 years research work in Donal's Lab, I developed some understanding about breast cancers. I still felt it was a big puzzle to me and I didn't have the answers about the cause and treatment. In the past 10 years, the breast cancer issues occupied both my mind and heart. I continue to work on the subject while I was working on jobs not directly related with breast cancer research.

After a few years of study, I finally feel I start to see the light and the answer becoming clearer to me. However, I feel disabled when I was trying to explain my understanding to anyone else. I am not generally a good communicator. Even after I attended 5 workshops and many coaching sessions for improving my communication skills, I still feel very difficult to explain my ideas to others. I decided to write a book. Partially because a book has an ample room for me to throw some free discussions and introduce ideas in a relaxed chatting fashion. In this book, I present to you some principles guided our body design and development. I hope those principles can help us to understand how we should take care of ourselves.

Part I
The Design of a Woman's Body

I was born in a small town, Weihai, China. I spend most of my life in Qingdao (Tsingtao) City. Both my grandmother and my mother graduated from American Catholic schools in Shanghai. I, however, don't believe that I belong to any religion. I consider all religions have good points. Strangely, since I was a child, I always feel that I was created for a purpose. I was wondering who created me. Because there were many coincidences in my life, I felt that some supper sprite was taking care of me. After read a huge Bible book written in Guangdongness, I found the story in the Bible was fascinating, sounds like a good fiction story.

One day, I changed my perspectives.

In 1989, I was studying in The Florida State University in Tallahassee. One day, I had lunch with two of my friends in the front yard. They were both Christians from Taiwan. It was a beautiful day. We were eating at a long table next to a giant tree. My friends told me their dating stories. It was a silence for a while. I saw an old man on the top of the tree.

The man wares a silver coat and cat. He has a lot of bear. His voice is low. He told me that I should not worry about my study.

Here is the story I heard:

In the quiet Universe with many naked planets, the LORD God, Yahweh Elohim, was bored. God made many little things: water, sugar, gas. A lot of things he made had interesting shapes, such as rings or strings.

Now God is excited because he is trying to make something complicated and more interesting. Ah, let me do a little experiment first. He then made some living things: from seaweeds to flowers and trees. These things were called as Plant. Leafs of these plants had different shapes but all of them had similar color. The color was similar to one of the God's favored stones - emerald. "Green color can comfort my eyes", God was glad. The flowers had different colors. Some of them looked like ruby and some of them emerald. In the beautiful garden, God was amazed that there are so many colorful flowers. After a while, God felt lonely in the beautiful garden. He decided to make some moving companions. He made birds, rabbits and monkeys. It is fun! God enjoyed watching them running around. God was not satisfied. These things cannot interact with ME. He decided to make another creature: Adam, an image of himself. This thing could worship me – God thought.

He made plants, from seaweeds to flowers and trees. He made some small animals, from flies to elephants. The human body was made of the element available in the universe: Some water, some soil, some metal and some gas. When things were mixed together, there comes Adam, a image of God.

Adam was lonely. Adam wasn't interested in being with any other animals made by God. God had already considered this in the initial design. He wanted to make a companion for Adam. God used an Adam's rib to make another creature: Eve. Oh, he created Eva with affection. When God made Eve, he did some modifications. Eve lost something a man had: a lot of body hair and the little thing. Eve had to be someone Adam love to be with all the time. God imagined the Being whom himself would like to share joy and life with. Here she was. Eva looks similar to Adam. However, Eva was a finer creature than Adam. Eve had smooth skin, soft body lines and sweet voice. Eva had something Adam didn't have: beautiful breasts on her chest.

God was a genius designer. Everything he did had purpose. Why should Adam and Eva close to each other? Yeah, God didn't want they live forever liked himself. God didn't want to make another God. 'These things are called Human Beings. They should not be immortal'. God thought. They must die after a while. It was a lot of work to make these things. Such complicated system and huge amount of details. God didn't want to make them again and again. These two things were the templates for the following generations. God built in the mechanism of replication into Adam and Eva's body. If Adam and Eve loved each other enough, they would know how to produce replicates. Apparently they did: here we are.

This was the thought of God:

In order to prevent human to rule the universe, these creature need to die after a certain number of years. How long should they live? 100 years is a good number. Adam does not know how to take care himself. Eve shall live a little longer to make sure that Adam has companion at the end of life. Ok, if Eve can live for a hundred years, Adam shall only live for 94 years. God is satisfied with the plan.

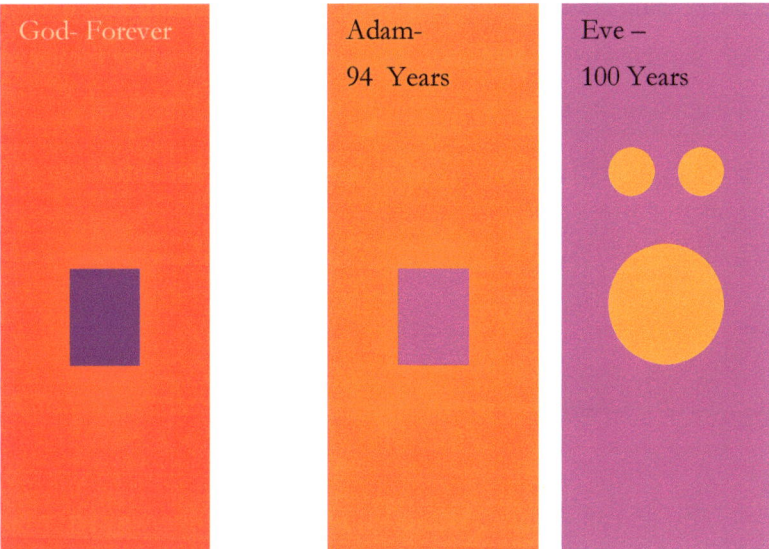

The next question is, how can they die?

This is a good question. They could kill each other. No. What if they don't want to? Sometimes they don't hate each other. It is much better to let the bodies die by themselves so that they will die for sure.

There are a few ways the body could die:
 i. The body will gradually become weaker and lost function after the whole body growing up;
 ii. Some tissue in the body will lose control in growth and these tissues can kill the whole body;
 iii. Some small things can go into bodies or mess up with the body.

How can I control the process? I can make some signals.

Why Breast Cancers?

When you look into a mirror, have you ever feel proud of yourself: "Oh, how beautiful I am!" Yes, you are so beautiful! Look at your eyes, they look charming and tender. Your skin is very smooth. The most attractive feature is that your body is lined with curves around your chest, hips and wrist. "God must made Eva with a lot of love." I often thought about this when I was a young child.

Later I realized that women are not only made look beautiful, they are also carrying heavy duty and enduring miseries. I remember that when I was about 8 years old, I couldn't stand seeing a neighbor lady got pregnant. Her adorable body became out of shape. I felt so sorry for her. She gave birth to a baby girl. Sometimes I saw her feeding the baby with her breast. She seemed not aware that others could see her breast. Only thing mattered to her was her baby. While I still felt pity for her losing her shape, I deeply admired her devotion to her baby.

Yes you are so beautiful on your appearance. To me, what makes you most beautiful is your kindness, your tenderness and your talents.

No matter how wonderful and beautiful we are, so many of us got breast cancers. Many of us die of the disease. I felt so confused.

I was convinced that God loves Eva and all women. However, I was wondering why HE would let us get breast cancers. I particularly don't understand why HE let us get breast cancer after we finished carrying and feeding babies. Why does HE punish us after we have done so many kind things?

My mind was restless. I almost felt upset about HIM. But I wanted an answer. There must be reasons why we get breast cancers. There must be an answer!

For years, I was seeking for the answers.

One night, I had a bite of chocolate. The second morning, I woke up earlier. I heard: "Yes, I love you. That is why I am keeping talking to you. I want you understand me, so that you can explain it to other women. I love you all."

Oh, God! I realized that is why we got breast cancers: we haven't understood HIM yet. I was lightened up. I felt that I find the clue about the answer. I believed HIM. HE must love us. He didn't design us to get ill. It is just like the car designers don't want the drivers to run into a car accident. Even if the car was designed to be safe, we might still likely to run into accidents anyway. If we know better about our cars and traffic rules, in addition to driving more carefully, many car accidents could be avoided. So is the situation of our lives. We need to understand HIM better. We need to know the rules the God used in designing Eva. If we know HIS rules and carefully follow the rules, we might be able to reduce the risk of breast cancers. Even if we got it, we might be able to treat it with ease.

God love us! After I realized this is true, I felt so much happier.

The question is: how did HE design us? I kept asking HIM.

"Think systematically!" HE said.

Our Body Systems

"System? What is that?"

This is a good question. There are many definitions about *System*. The common line of words is the following:

A System is a set of physical parts or organization groups that interacting and collaborate each other by following certain rules in order to achieve a common goal.

Some good examples of systems are computer systems, network systems, social system and school systems. You might not realize how many systems you are dealing with every day.

An example that may familiar to us is the designing of a car. A car has an engine, a brake and a transmission and many other subsystems. Each of these subsystems is essential for a car to run. At the mean time, they have to work collaboratively. If they don't work together, even if each part works perfectly, if they don't work with other subsystems, or one of the sub-system is broken, the car would be dead.

Our interest is on ourselves, mainly our body. You must have noticed that our bodies have many different parts. Women have several special types of parts that are different from men. For example, obviously, we have much bigger breasts than those of men's because men's breasts are not developed.

There are so many parts in our bodies to list. We might often concern about the well being of each part of your body. However, the God didn't make us as a mechanic assembly of many parts. Instead, HE figured out the best way to make these parts to position and work together. Remember that God is a genius who can do great work in all aspects of creation. HE designed Eva's body not only in the most beautiful shape for her to look beautiful, but also as a most intriguing, efficient and complicated super-system in order for her to live well. This super-system contains many subsystems. Each subsystem is working independently to perform certain function. All subsystems are collaborating with other subsystems to make our whole body functional. For example, in our body, there is a digestion system to help us to digest food, a respiratory system to help us to breathe fresh air. Both these systems are aimed at helping our body alive.

Ever since I started to study biology, I have been constantly feeling amazed about how genius the God's design is. Sometimes I really doubt that the God can design such a wonderful system in such a great detail all on HIS own. "HE must have hired some very talented architects and army of smart engenderers." I often guess.

Anyway, let's not to concern about how the God did HIS job. It is more important to take care about ourselves since we have been fortunately created by HIM.

Remember that the God made Adam and Eva separately. As you know, at first HE used mud and made Adam. HE then used one of the Adam's ribs and made Eva. Therefore, there are some significant differences between Adam and Eva's bodies. All of men and women are human beings, regardless our race. We all need to eat and breathe. However, men and women are playing different roles in making children. Therefore, men and women have different reproduction systems.

There are three major systems in women's bodies are significantly different from those of men's:

- **Reproductive system** – the system to produce babies;
- **Baby feeding system** – the system to feed babies;
- **Endocrine system** – the system to regulate the reproduction and baby feeding systems.

In order to understand ourselves, it is very important to view our body in a systematical way so that we can understand how the God designed us.

Women's Reproduction System

Although both men and women are needed to produce children, the God wanted women carry more responsibilities in producing and breeding children. The reproductive system in women is so heavily loaded with the following large components:

1. Ovary;
2. Fallopian tubes
3. Uterus (womb);
4. Cervix;
5. Vagina (birth canal).

We have uterus and ovary. They allow us to create, carry and feed a baby. However, they are very vulnerable and easily get ill.

Baby Feeding System

Breasts are the center of our beauty and kindness. We are using them to feed our babies.

The breasts contain mammary tissues and glands for the production of milk for our babies.

Women's Endocrine System

How could the breasts know what happened in the ovary? It looks like some organization is telling each of our body what to do and when to do it. What is it?

It is the endocrine system that produces signals to regulate all aspects of our bodies. The signals are the hormones we mentioned above. In this book, I will only discuss about the part of endocrine system associated with breast cancers.

The following parts of our body are associated with the reproductive and baby-feeding related endocrine system:

1. Hypothalamus in brain;
2. Anterior pituitary in brain
3. Ovary;
4. Uterus
5. Breasts

   Many other parts of our bodies are also related with our endocrine system:

   1. Adipose
   2. Muscle
   3. Bone
   4. Liver

The Female Hormone

Our body has so many different parts. Have you ever wondering how these parts grow and work together? For example, ever since we were about 12-15 years of age, we started to have monthly period.. The normal interval between 2 periods is 28 days. Sometimes external stimulus might interrupt the timing. Unless there is there is some problem, most periods are supposed to be regular. Along with the period progressing, we started development of our body. Our body shape started to show more curves and breasts started to be bigger. The worst of all, we started to be interested in boys! Another example of our body parts work together is that right before a lady gives a birth to a baby, her breasts starts to produce milk for feeding the baby. It seems that our body parts are talking to each other. There must be something coordinates different parts of our body.

You are right. This thing is called *Hormone,* a Greek word means "to set in motion". Actually there are many hormones in our bodies to organize and facilitate our body parts to communicate among them. The most important hormone for women is *estrogen.* The most important hormone for men is *androgen.* Interestingly, all estrogen is made from androgen. Remember that the Bible told us that Eva was made from a rib of Adam. I am not sure whether that story was true. However, it seems true that women are made from men's hormone.

You are right. This thing is called *Hormone,* a Greek word means "to set in motion". Actually there are many hormones in our bodies to organize and facilitate our body parts to communicate among them. The most important hormone for women is *estrogen.* The most important hormone for men is *androgen.* Interestingly, all estrogen is made from androgen. Remember that the Bible told us that Eva was made from a rib of Adam. I am not sure whether that story was true. However, it is true that women's hormones are made from men's hormones.

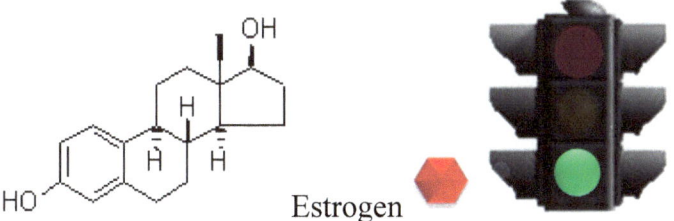
Estrogen

There are three types of estrogens:

1. Estriol

Estriol is produced during pregnancy in the normal human fetus. The production of this type of estrogen goes through our life time. Men also have this type of hormone.

2. Estradiol

This form of hormone is strong and produced in ovary when women between the first menarche and menopause. This type of estrogen makes women distinguish from men, the hormone that is specially designed for young women.

3. Estron

This form of hormone is strong and produced in non-ovary tissues after women having menopause

In most cases estrogen wants thing to grow. If estrogen is allowed to make all the decisions, everything will keep growing and out of control. A good thing is that our body is also making another hormone called progesterone that can keep estrogen in check so that estrogen can only have limited amount of power.

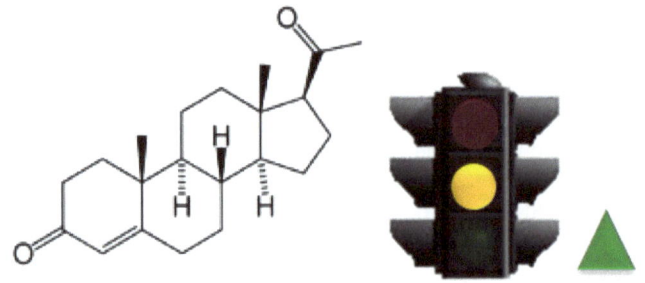

Before menopause, both estrogen and progesterone are synthesized in either ovary or adrenal gland in the brain. After menopause, most of them are originated from the adrenal gland.

Total Estrogen Equation

Each part of our body is affected by total estrogen at that area. For each part of body, the estrogen concentration is different dependent on tissue, time and age.

I assume that God didn't go to school to learn calculus. All God knows is the logic and simple calculations. The following the total equivalent estrogen in each part of our body:

[Estrogen] = A * [Esdiol] (tissue, pregnancy) + B * [Estradiol] (women, tissue, time, menoage) + C * [Estron] (women, postmeno)

Here A, B and C are the efficiencies for reflecting the strength of each type of estrogen.

Over our life time, the types of the estrogen in our body are changing. The following graph showing the estrogen types in our bodies at different ages:

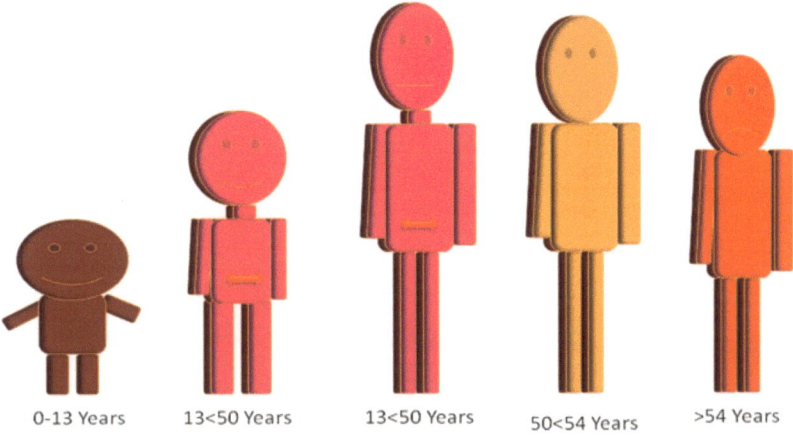

0-13 Years   13<50 Years   13<50 Years   50<54 Years   >54 Years

The following chart showing you how each type of estrogen contributes to the total estrogen amount in our bodies in our life time:

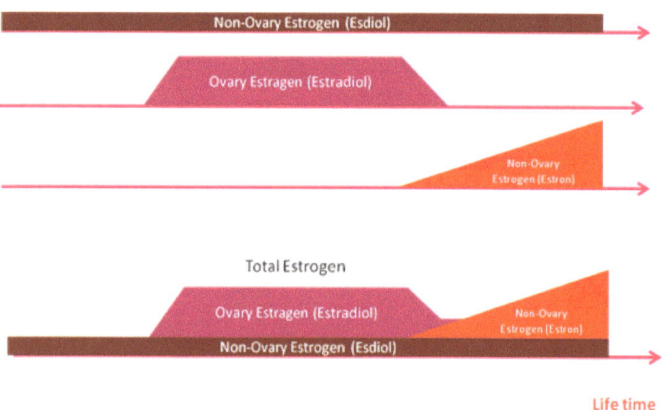

From the above chart, you can see that the concentration of estrogen in our body is not remaining the same level.

The total level of estrogen is like this:

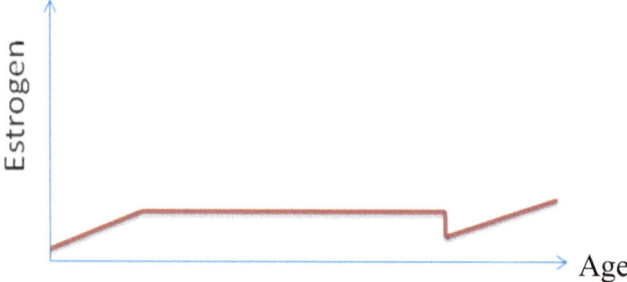

Estrogen Receptor

Estrogen binds to estrogen receptor to work. If there is no estrogen receptor, estrogen is useless. Estrogen is like a car key. The Estrogen Receptor is like a car engine. A cell is like a car. A car will not run without key turns on the engine.

Total Estrogen Sensitivity

Estrogen binds to estrogen receptor to work. If there is no estrogen receptor, estrogen is useless. More estrogen receptor in a cell, the more sensitive a cell will response to the estrogen activity.

As women age, the amount of estrogen receptor will increase to compensate the loss of estrogen.

In most cases estrogen wants thing to grow. If estrogen is allowed to make all the decisions, everything will keep growing and out of control. A good thing is that our body is also making another hormone called progesterone that can keep estrogen in check so that estrogen can only have limited amount of power.

As we age, the level of progesterone decreases too. The rate of decrease is faster than that of estrogen. Before menopause, both estrogen and progesterone are synthesized in either ovary or adrenal gland in the brain. After menopause, most of them are originated from the adrenal gland.

Therefore, the estrogen sensitivity of a cell can be described as the following equation:

[Estrogen ⬢ Sensitivity]

= [Total Estrogen ⬢] X [Estrogen Receptor] -[Progesterone] X [Progesterone Receptor]

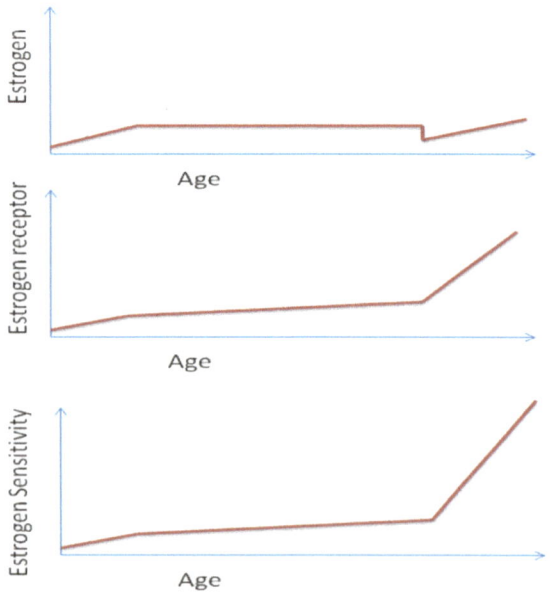

It is interesting to compare the chart we showed in the beginning of the book:

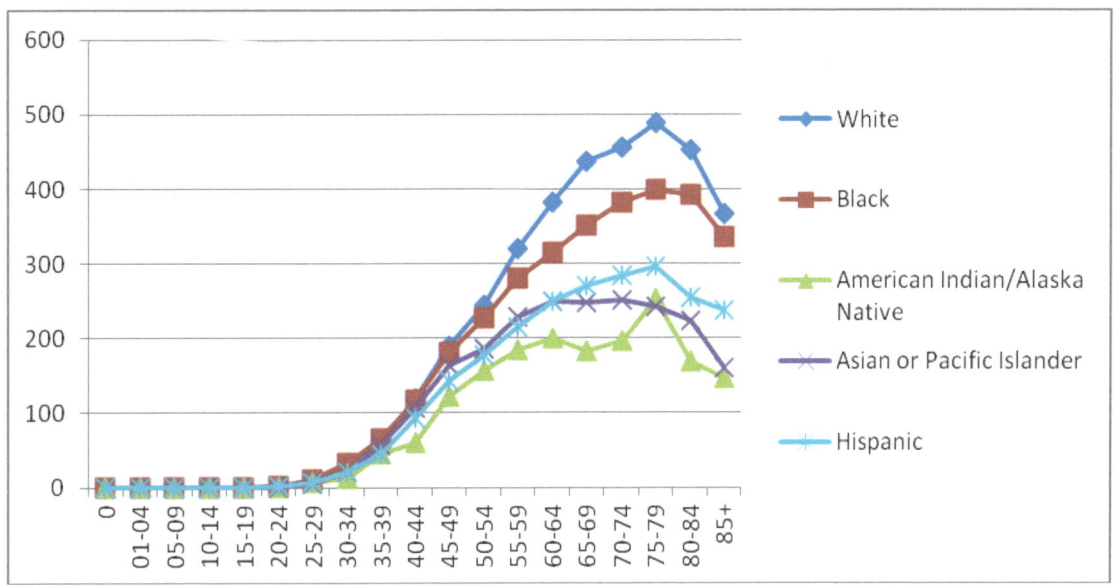

This chart confirmed that both the toltal estrogen sensitivity might be the reason that women got breast cancers.

Women's Endocrine System

How could the breasts know what happened in the ovary? It looks like some organization is telling each of our body what to do and when to do it. What is it?

It is the endocrine system that produces signals to regulate all aspects of our bodies. The signals are the hormones we mentioned above. In this book, I will only discuss about the part of endocrine system associated with breast cancers.

The female endocrine systems change during our lifetime. The following parts of our body are associated with the reproductive and baby-feeding related endocrine system:

6. Hypothalamus in brain;
7. Anterior pituitary in brain
8. Ovary;
9. Uterus
10. Breasts

The following process showed the process that hormones are made:

Cholesterol -> Pregnenolone -> Progesterone -> Aldosterone -> Testosterone -> Estradiol

The pituitary gland in a brain can produce a hormone called LH (Luteinizing hormone). It will travel to ovary to stimulate the ovary releasing the female hormone, Estrogen. Estrogen will be secreted into blood vessels and move to all over the body. Estrogen then goes into breast tissues to stimulate breast growth.

When you are aware your appearance, I am wondering whether you ever think about your own body. I don't mean how beautiful your body is, I mean what is your body made of and how is it made. I am pretty sure that some of you are already very knowledgeable about your body. However, someone else may not. I would like to bring everyone on to the same page. Therefore, I will remind us some basics about our body. Here I will only mention the parts of a body, which may relate to the disease mentioned in this book.

The Brain

Everyone knows that each of us has a brain. The reason I mention this is that a brain plays a very important role in many diseases, including breast cancer. The brain I mentioned here is a basic brain. It doesn't matter how intelligent or naïve you are, the basic women's brains have same structure and function. What is in the brain? There is a central neuron system that reside our conscience, feelings and thoughts.

When I was little, I was taught that my feeling was in my heart. I felt very uncomfortable when I learned later that, instead of being in the heart, our emotion is the reaction in our brain. It was so hard to believe. Sadly, it is true.

In the brain, there is a part called pituitary gland. Please remember this gland. This part can produce a hormone called LH (Luteinizing hormone). Later we will see how important it is for our lives.

When the tissue keeps constantly growing and shrinking periodically, you might not be surprised why breast cancer occurred so easily. It has been reported that women have later menopause will have higher risk of breast cancers.

Hopefully, by now you can have a simple picture about what is special in our women's body.

Part II
The Design of a Human Cell

A Basic Human Cell

Our body is made of many, many cells. How many? It is hard to say. Someone reported 50 million and others said 100 trillion. I know you are going to say, "Wow!" We can imagine how difficult it is to count the number of cells in our bodies.

Have you ever wonder about how the human cells look like? However, a human cell is very small. Unless you have a microscope, it is very hard to see a real human cell. A simple way to show you a human cell structure is to look at a chicken egg. There is a yolk surrounded with egg white. A basic human cell looks very similar to a chicken egg. We call the egg yolk as nuclei and egg white as cytoplasm.

The Cell Types of Human Cells

The cells are not the all the same in our bodies. There are germ cells made from stem cells. Sperm are made in testis and eggs are made in ovary. The organs in our bodies are built with somatic cells. Each organ in our body has a few special cell types. For example, the cells in a brain are very different from the cells in ovary. Even inside of brain, the cells are different in different section of the brain.

The Origin of Human Tissue Cells

Chicken or egg, which came first? This is an ancient question. I won't argue with you on this topic. However, I do want to discuss with you about how an egg becomes a chicken.

There was an egg in a mother's uterus made in ovary. After received a sperm, the egg started to divide into two cells.

http://images.search.yahoo.com/search/images/view?back=http%3A%2F%2Fimages.search.yahoo.com%2Fsearch%2Fimages%3Fp%3Degg%252C%2Bhuman%26ei%3DUTF-8%26x%3Dwrt&w=501&h=482&imgurl=www.sp.uconn.edu%2F%7Ebi107vc%2Fimages%2Fcell%2Fsperm%2Begg.JPG&rurl=http%3A%2F%2Fwww.sp.uconn.edu%2F%7Ebi107vc%2Ffa02%2Fterry%2FLect1.html&size=43.0kB&name=sperm%2Begg.JPG&p=egg%2C+human&type=jpeg&no=3&tt=1,600&oid=21485ce68b5d9450&ei=UTF-8

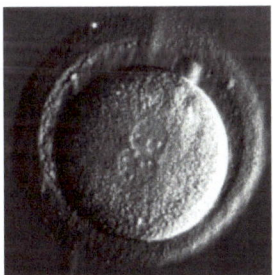

http://www.advancedfertility.com/pics/zygoye.jpg

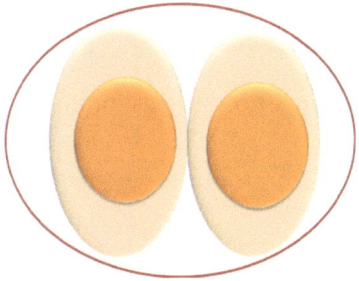

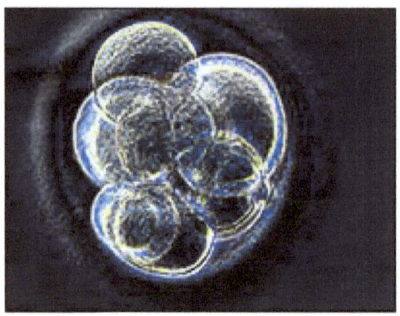

http://images.search.yahoo.com/search/images/view?back=http%3A%2F%2Fimages.search.yahoo.com%2Fsearch%2Fimages%3Fp%3Dembryo%26ei%3DUTF-8%26fr%3Dyfp-t-501%26b%3D121&w=203&h=152&imgurl=newsimg.bbc.co.uk%2Fmedia%2Fimages%2F41167000%2Fjpg%2F_41167193_embryo203.jpg&rurl=http%3A%2F%2Fnews.bbc.co.uk%2F1%2Fhi%2Fhealth%2F4563607.stm&size=7.9kB&name=_41167193_embryo203.jpg&p=embryo&type=jpeg&no=138&tt=78,984&oid=69595d71bcbebaf4&ei=UTF-8

http://images.search.yahoo.com/search/images/view?back=http%3A%2F%2Fimages.search.yahoo.com%2Fsearch%2Fimages%3Fp%3Dcell%252C%2Bdivide%252C%2B%26ei%3DUTF-8%26b%3D61&w=288&h=288&imgurl=www.advancedfertility.com%2Fpics%2F8cellicsi.jpg&rurl=http%3A%2F%2Fwww.advancedfertility.com%2Fembryoquality.htm&size=11.2kB&name=8cellicsi.jpg&p=cell%2C+divide%2C&type=jpeg&no=64&tt=564&oid=dbd4ae6e9d5cb12a&ei=UTF-8

You must have noticed that, up to the morula stage, the cells are round and evenly distributed in the embryo. These cells are very similar to each other. They are not yet specialized. If we separate the embryo at the eight cells stage, each of these cells can grow into eight individual embryos. We can repeat this process over and over again. Eventually, we can make more than a million embryos from one fertilized egg.

After morula stage, the embryo will develop into blastocyst. The cells will become more and more differentiated depending on where the cells are located. It seems that the cells start to talk to each other. The differentiated cells cannot grow into embryo any more. However, a group of cells inside of the blastocyst still can grow into embryos. These cells are called stem cells.

The Lifetime of Human Cells

Have you ever desired to live forever? I do sometimes. I gave up lately because I realized that it is not possible. However, believe or not, embryo cells can. Just like I mentioned before, the embryo cells or stem cells can keep dividing forever. Why we are still dying? Good question! That is because that the somatic cells can only divide about 50 times. The will die after wards.

We were taught that our ancestor, Adam and Eva were made by the God. However, we know that we were made by our parents, father and mother. What they did is that the father donated a sperm and the mother contributes an unfertilized egg:

How are different a normal cell a cancer cell different? A living cell is like wheel going downwards.

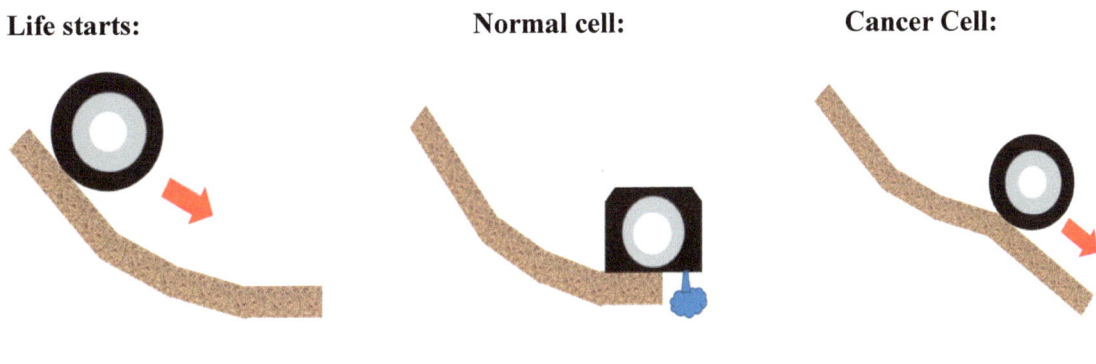

Part II

The Key Players in a Human Cell

Proteins are the Players

From the previous discussion, you might have an impression that a cell is complicated. . I thought so too. The design of a cell is ingenious. I was beating the table while I had just understood the mechanism. I feel that I saw the light: God is a genus! As a non-religious person, I couldn't help to admiring His design.

The questions are: How are the cells controlled or regulated in such an organized fashion? Who are the players, regulators and general commanders?

The answer is not simple: There are many other types of molecules in a cell, such as hormones, DNA, RNA and small molecules. Most cases, proteins are the players, regulators and general commanders. I will discuss about these types of molecules when we come along.

Protein Families

I am wondering whether you have a big family. I believe that everyone has a father and a mother, except Jesus. Do you also have brothers or sisters? I have two younger brothers and, unfortunately, no sisters. I have a son. My parents have many relatives. Therefore, I have many uncles, aunts and cousins.

Members of the nuclear family use descriptive kinship terms:

- **Mother**: a female parent
- **Father**: a male parent
- **Son**: a male child of the parent(s)
- **Daughter**: a female child of the parent(s)
- **Brother**: a male child of the same parent(s)
- **Sister**: a female child of the parent(s)
- Grandparent
    - **Grandfather**: a parent's father
    - **Grandmother**: a parent's mother
- **Grandson**: a child's son
- **Granddaughter**: a child's daughter

For collateral relatives, more classificatory terms come into play, terms that do not build on the terms used within the nuclear family:

- **Uncle**: father's brother, father's sister's husband, mother's brother, mother's sister's husband
- **Aunt**: father's sister, father's brother's wife, mother's sister, mother's brother's wife
- **Nephew**: sister's son, brother's son
- **Niece**: sister's daughter, brother's daughter

Cousin: the most classificatory term; the children of aunts or uncles. Hence the phrase "third

In most culture around the world, ever since we entered so called human civilization, a family has been the core unit of a society.

What does a family mean to us? First all, the father and the mother have to meet in order to produce a child. The father and the mother might not have to marry. They might be just dating. I am talking about the nature way to produce a child, not the artificial approaches such as in vitro fertilization or cloning. A child is carrying the combined genetic features from the parents. Sometimes, the birth parents may not live together anymore. The child might be adopted by the step parents. As an increasing trend, many children have been adopted by their parents who are not their birth parents at all.

People in a family might love each other, sometimes compete with each other and sometimes even hate each other. There are a lot of behavior similarities in a family due to either genetic reason or the common living environment they live with for years. At the mean time, very rarely, any two members in a family are doing the same things all the time, even they are identical twins. Very likely, each member of the family works takes his/hers part of responsibility to achieve the common goal of the family.

A cell is like a world. Proteins in a cell work very similar to people in our society. Each protein belongs to a family and the protein families are the core units in a cell

I will introduce a few cancer related protein family to you so that you can feel comfortable later when I talk about them again later.

P53 family

P53   P51A   P73B

http://images.search.yahoo.com/search/images/view?back=http%3A%2F%2Fimages.search.yahoo.com%2Fsearch%2Fimages%3Fei%3DUTF-8%26p%3Dp53%252C%2520family%26fr2%3Dtab-web%26fr%3Dyfp-t-501&w=720&h=540&imgurl=www.idac.tohoku.ac.jp%2Fdep%2Fcellbio%2Fp53%2520family%2F53family.jpg&rurl=http%3A%2F%2Fwww.idac.tohoku.ac.jp%2Fdep%2Fcellbio%2Fp53%2520family%2F1.html&size=40.6kB&name=53family.jpg&p=p53%2C+family&type=jpeg&no=17&tt=77&oid=1ca42cdada56db64&ei=UTF-8

Based on many reaserchers publicated results, the following graph showed how P53 drive cell death or growth.

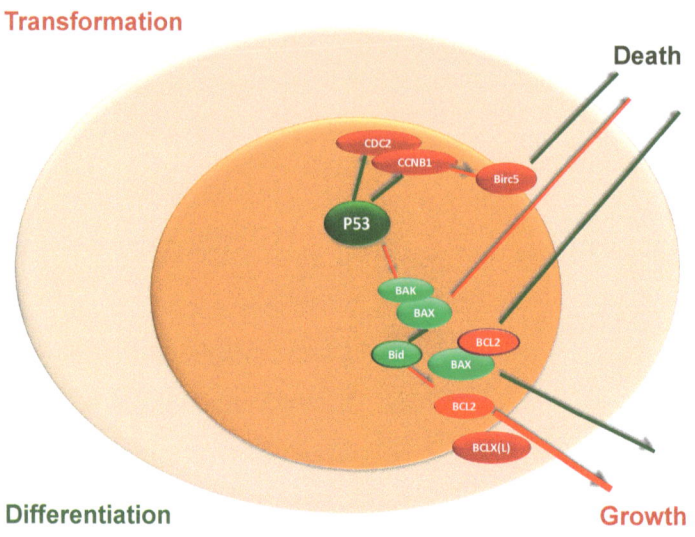

Myb Family

Unlike P53, Myb family is not famous. However, Myb family is one of the most important protein families in cancer origination and development.

There are 3 Myb family members:

Myb,

MybL1 or AMyb

and MybL2 or BMyb

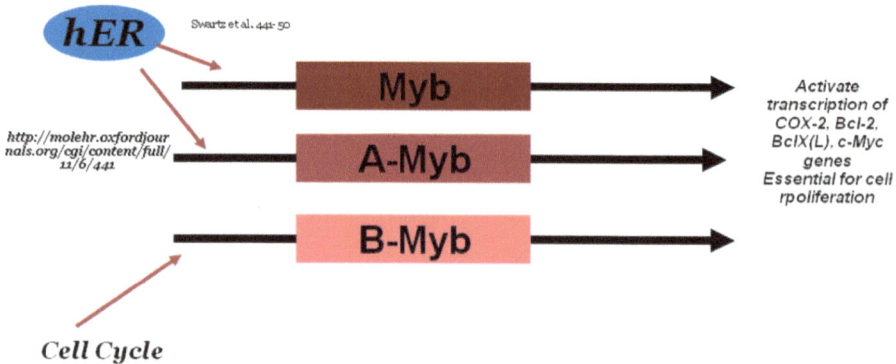

The total Myb effect is:

Total Myb = Myb+ MybL1 (AMyb)+ MybL2 (BMyb)

Estrogen Receptor Family

From Wikipedia, the free encyclopedia

*There are two kind of **estrogen receptor** (ER): ER Alpha and ER Beta.*

P53 and Estrogen Receptor Rule the World

Part III

The Mechanisms of Breast Cancers

The Origin of Cancers

The human breast cancer claims more than 100,000 women's lives per year world wide and 1 out of 7 women have breast cancer at different stages in the United States. To understand the mechanisms and development of breast cancer at molecular and cellular level is urgently important. The genetic relationship among breast diseases includes fibrocys diseases, hypertrophy and cancer. The breast cancer is an accumulative progressing result that started from fibrocys diseases to hypertrophy, and end with cancer:

Normal breast -> fibrocys diseases -> Hypertrophy -> Breast cancer

Genetic events occurred in each disease, which are associated with multiple signaling and metabolic pathways. Our results also showed that PPARG might play an important role in the development of breast cancer.

Breast Cancer Types

There are two breast cancers types:

1. Somatic Cancer – ER positive cancer
2. Embryonic Cancer – ER negative cancer;

Somatic Cancer – ER positive cancer

Over 70% of breast cancer patients have detectable levels of the human estrogen receptor (hER) in the breast tissues (hER positive), while rest of breast cancer patients don't have detectable hER in the breast tissues (hER negative). The fact that anti-hormone therapies are effective for treatment of hER positive patients indicates that estrogen receptor activity is involved in breast cancer occurrence. In addition, the survey conducted by NIH showed that the women who are receiving hormone therapy are more likely to develop breast cancer in later time. This evidence indicates that estrogen receptor might be the driving force for caner to occur.

Embryonic Cancer – ER negative cancer;

It is much more difficult to treat the breast cancer patients who don't have detectable estrogen receptor in their breast tissues. These hER negative patients don't respond to anti-hormone therapy. It is less clear how the breast tissue cells become cancerous without estrogen activity

It was shown that women with P53 mutations will likely develop breast cancer in their life time. P53 is playing a very important role in regulating cell growth and programmed cell death through controlling multiple pathways. Although we understand how P53 regulate each pathway, it was not clear how P53 and the human estrogen receptor work together to determine the cell fate, such as cell growth or cell death. Targeting P53 has proven to be not effective because P53 involved in so many aspects of cell life so that targeting P53 causes many undesired cellular effect. If we can determine a minimal pathway network that determines cell fate but only associated with smallest number of proteins, it may allows us to identify targets that can produce desired cellular effect for cancer treatment. Therefore, it is very important to systematically understand the breast cancer cellular pathway network for both estrogen receptor positive and negative breast cancer cells.

One of the most significant features of early cancer cells is that the cells are undergoing proliferation without proper control. Proliferation means that cell growth rate is faster than cell death rate so that cell number will increase. Proliferation in solid tissues may cause a local benign tumor growth. However if the cell lost ability of responding to any extracellular signals and cell-cell communications for growth and death, the cells are transformed to be malignant.

In this work was aimed to construct a minimal cellular model to explain how a breast cell can become cancerous. A cellar pathway model was developed based on public understanding on large numbers of regulatory proteins.

Construction of the Minimal Cellular Pathway Model for Breast Cells

The cancer cells have the following possible cell fates:

1. **Cell death** (opposite to cell survival) is regulated apoptosis pathway and cell survival pathway ;
2. **Cell growth** (opposite to cell cycle arrest) is regulated by cell cycle pathways;
3. **Differentiation** is driven by cell differentiated growth pathway;
4. **Transformation** is regulated by cell transformation growth factor pathway;

The scheme of the cell fates:

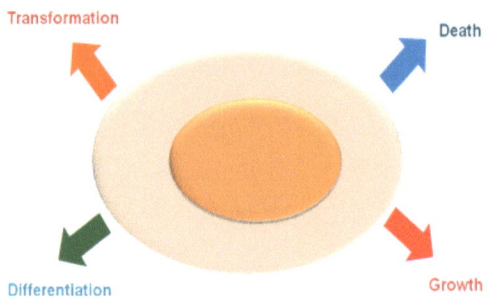

Different cell types showed different cell fate

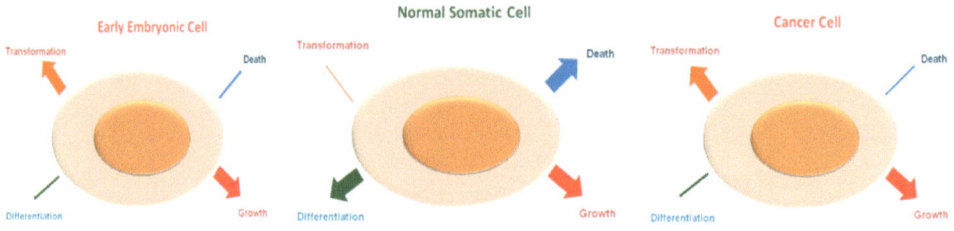

After all, cancer cells are so much like early embryonic cells.

In this model, the cell proliferation rate equal to that of the difference of cell growth and cell death:

Proliferation = Growh – Death (1)

The degree of malignacy can be measured by disfference between the degrees of transformation and differentialtion:

Malignancy = Transformation – Differentiation (2)

The cell will be cancerous if it is positive on both proliferation and malignancy.

Cancerous = (Proliferation >0) & (Malignancy>0) (3)

Identifying cell cycle regulatory proteins directly regulated by hER

Several publications reported that hER activates expression of c-Myb and MybL1 (AMyb) genes . Myb protein family contains Myb, MybL1 (AMyb) and MybL2 (BMyb). While Myb, MybL1 are transcriptional activated by hER, MybL2 is regulated by cell cycle proteins instead of by hER. Interestingly, all these 3 proteins are binding to the same DNA sequence and target the same genes.

The following picture showed the cell growth and apoptosis pathways regulated by hER:

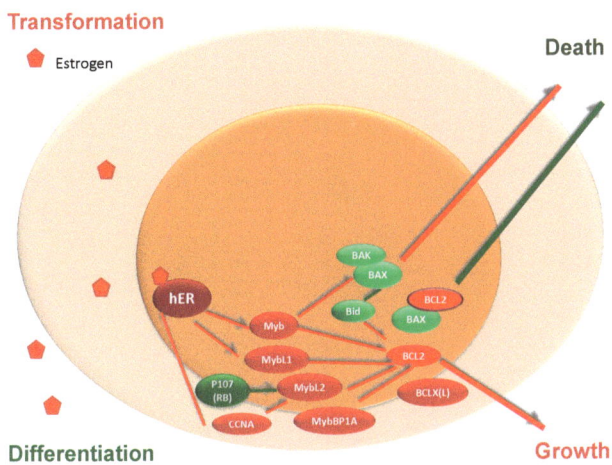

Identifying cell cycle regulatory proteins directly regulated by P53

Many proteins have been shown to be regulated by P53. We are only interested in proteins that have the most direct impact on cell growth and cell death.

P53 is playing a commanding role in inducing apoptosis. Mitochondria are believed to be the central stage in p53-mediated apoptosis. However, the signal transduction pathways leading to mitochondria remain unclear. It is report that P53 up regulate the translocation of Bax protein from cytosol to mitochondria, and this process is required for p53-induced apoptosis. P53 binds to Bcl2 to prevent Bcl2 and Bax binding so that Bax and Bak can form dimmer for proceding apoptosis process.

In order to view the overall picture, we jointed the P53 and hER cellular pathways together. It showed the cell growth and apoptosis pathway regulated by hER and P53:

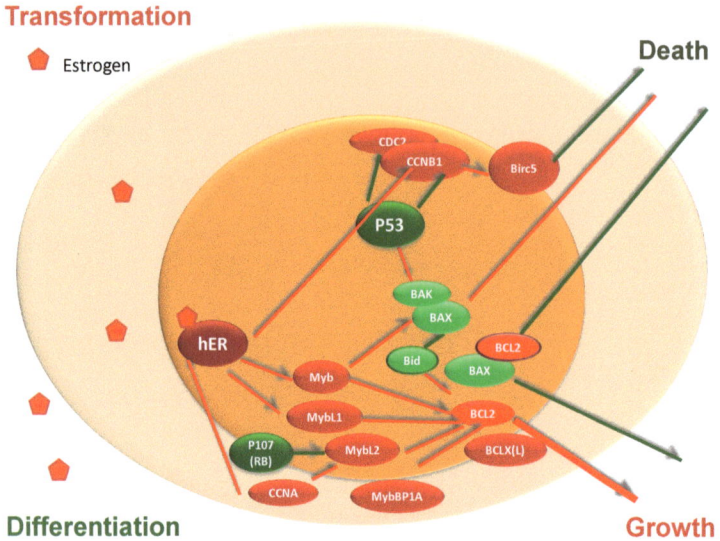

This model indicated that the cell fate is dependent on the balance between the rated of cell growth (largely depended on the effects of Bcl2 and BclX(L)) and the rate of cell death (determined by the activity of Birc5). While birc5 dependants on the dimmer of CCNB1/CDC2 activation, Bcl2 and BclX (L) are activated by the accumulative activity of Myb, MybL1 and MybL2.

The reason that Myb can cause cell dividing is that Myb activates oncogene C-Myc to drive cell divide.

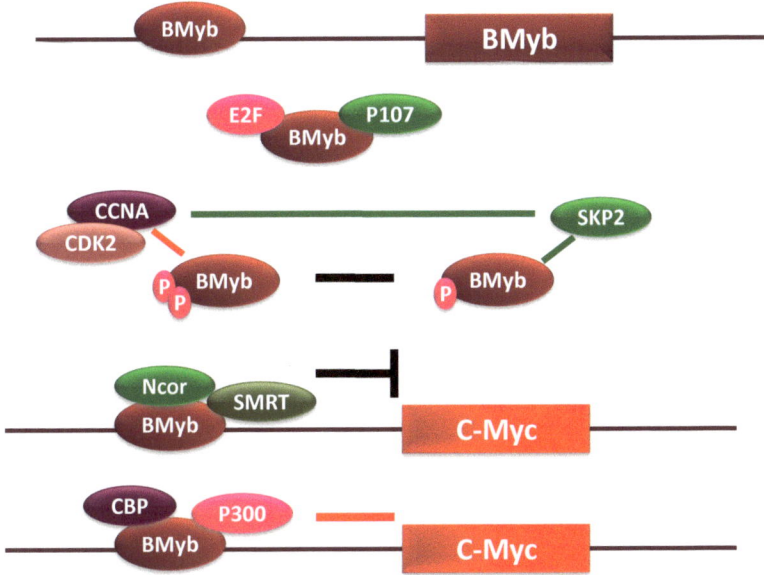

To understand how a normal somatic cell can be converted into a cancer cell, it is interesting to compare the normal somatic cells and cancer cells with embryonic cells. An early embryonic cell doesn't have tissue specificity. Its growth is programmed based on the genomic make up and content in the cell including proteins and small molecules. A normal somatic cell is a tissue specific cell and the rates of growth and death are partially controlled by external signals and cell-cell communications. The fate of a normal somatic cell is also controlled by internal signals, such as DNA damage or native concentration or activity level of certain proteins. It was well known for a very long time that a normal somatic cell can only divide a limited number of times before its death, while an early embryo cell doesn't have this limitation. The remarkable discovery a few years found out that the length of the telomere determines the number of times that a cell could divide. If in a cell the telomere is not shortening after each cell cycle, the cell will be immortal; this means that the cell can divide unlimited number of times. However, these cells might still respond to external signals for growth and death, as well as differentiation. After a cell lost its ability to respond to external signals, the cell will lose tissue specificity. If this cell proliferates, cell thus developed into a cancer cell. The cancer cells undergo continues proliferation with faster growing rate than cell death rate. After the cell completely lost tissue specificity, the cell will be able to move to other tissues and induce proliferation in the invaded tissues. Cancer cells are immortal and can continually divide without limitation of cell cycle numbers that usually applied to normal cells. Recall the nature of embryo cells, a cancer cell seems more like an embryo cell rather than a somatic cell.

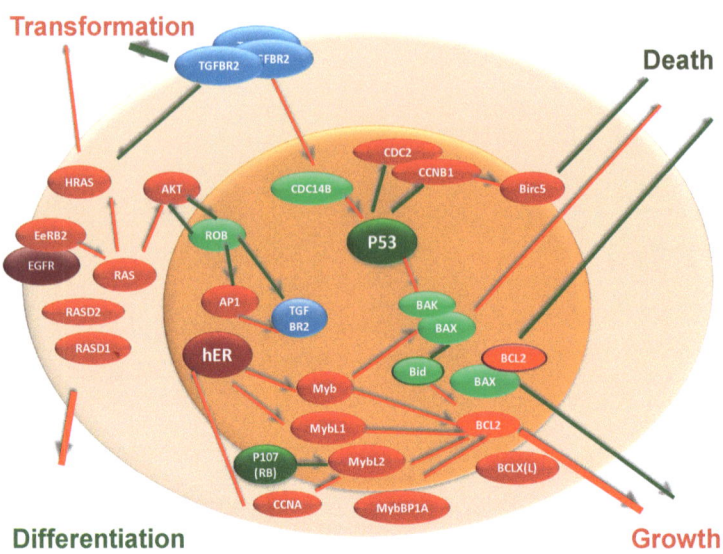

The following is the more complete network map of breast cancer tissue:

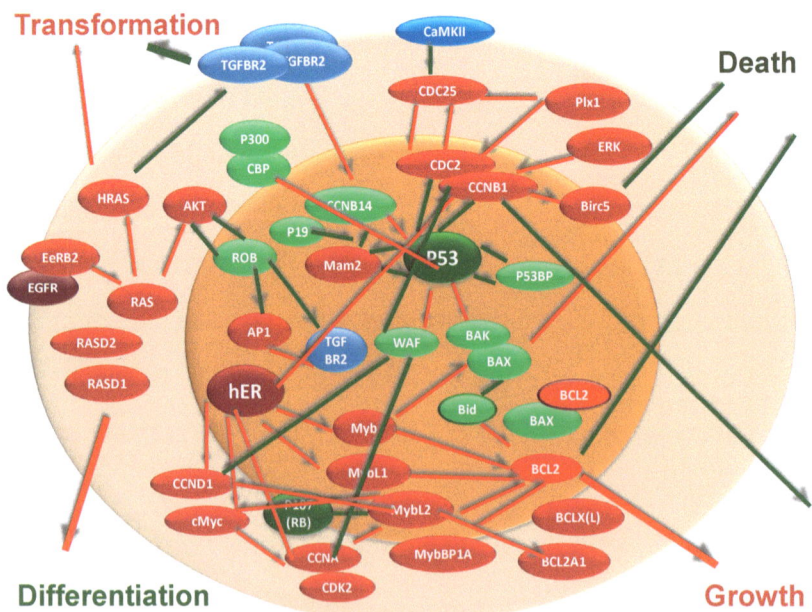

1. In the normal cells, cell death and cell growth are balanced so that the number of cells is constant;
2. In the normal hER positive cells, if estrogen level increases, hER will enhance transcription of Myb and MybL1, so that the rate of the cell growth increase. At the mean time, At the mean time, hER also induced CCNB1 expression, so that the rate of cell Embryonic or Somatic Cancer Types: How do CCNB1, CCNA, BIRC5, Myb, MybL1 and MybL2 Mediate Estrogen Receptor (hER) and P53 Controls of Cell Fate in Cancer Cell.

Part IV

Life Style and Breast Cancers

Reduce Stress

Everyone knows that stress is harmful for our health. Some survey showed that stress can promote tumor growth. However, we don't exactly know whether this is true and if so, why.

Twenty years ago, while I was in China, I heard a story from my adorable English Professor, one of the most talented and humorous American man I have ever seen. In an American History class, he told us a social study case on how marriage rate relates to war times.

Here is the story he told me:

It was reported by some survey that more people got married during the war times. A group of social scientists want to know the reason to explain the relationship of marriage rate with war times. They were trying to find out whether people tend to get married during the wars. One guess was that may be stress would make people feel love. The scientists wanted to validate this hypothesis. They wanted to see whether stress can affect people's emotional state. You can imagine that it was very difficult to simulate war in order to do a test for this hypothesis. They thus conducted an experiment in a mountain. The samples were two equally divided groups with same number of single men and single women in each group. The control group stayed in a beautiful area with flowers and trees. The hills slopes in the area were moderate so that it was pleasant. It was a paradise. The test group, however, were put in the area with steeper hills and difficult walking passes. The scenery had great sense of danger. It looks horrible!

After 6 months, the experiment was finished. The people in the control group became very good friends. They kissed and hugged each other and promised to visit each other in the future. Then they said good bye. However, the people in test group didn't make friends. Instead, they were either decided to get married or they will keep dating and make marriage decision in the near future.

What did this result tell us? This result confirmed the hypothesis that stress might help people fall in love and develop much stronger desire to have relationship with a person with opposite sex.

This story was hanging in my mind for many years. I didn't understand why stress has anything to do with love.

After I spent years in breast cancer research, I felt that I finally have better understanding about the mechanism behind of this phenomenon.

Here is what I believe have happened:

Remember the time when we were younger than 12 years old, we were not that much interested in boys. At least I didn't. I thought that boys were naughty, dirty and stupid. Occasionally I met some boys who were not that bad, but not enough for me to be interested in being with them. Only after we entered teenager age, we started to pay attention to boys. The boys became cute, handsome and smart in our eyes.

What was happening? What had been changed?

The boys were still same boys. Only we, the girls, changed. It was not our decision to change our selves. It was our bodies kept changing responding to increased estrogen concentration in our body. Eventually, this change reached a threshold. Not only we started to grow our breasts and have monthly periods, but also this change affected our minds too.

What does this story have anything to do with stress and breast cancer? Here is the logical hypothesis:

If romantic love between two opposite sex parties is driven by estrogen, and if stress can enhance the romantic love, then estrogen and stress might have some relationship.

However, we are not clear what the relationships are between two factors.

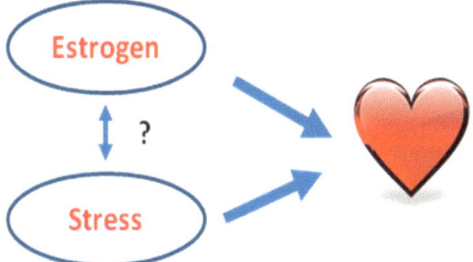

There are some results from a few studies have been reported:

1. Premenopausal stress will reduce estrogen **production in ovary**;
2. Estrogen is able to relief stress caused artery and heart problems;
3. Stress with high level of estrogen will cause more brain damage;

I have been very much bothered by this puzzle. May be this is the answer:

1. Nothing happened

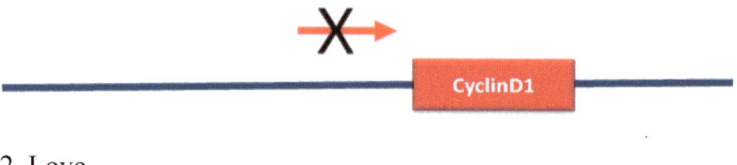

2. Love

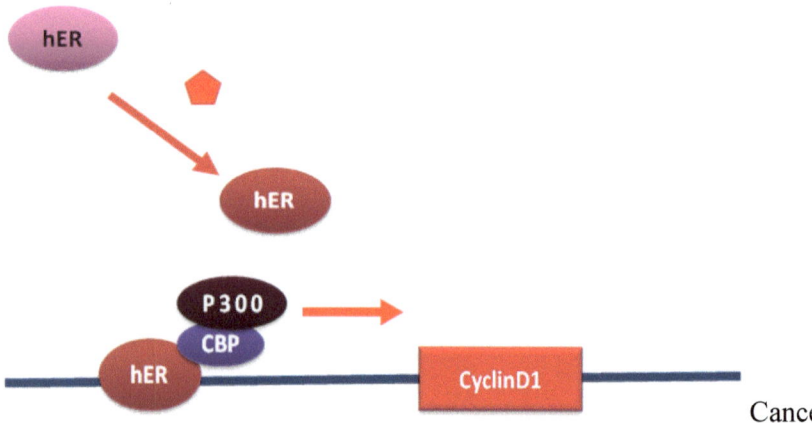

Cancer

3. Stress

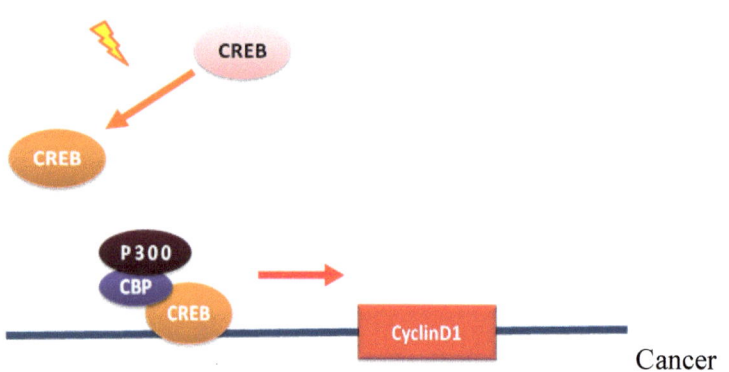

Cancer

4. Love and Stress

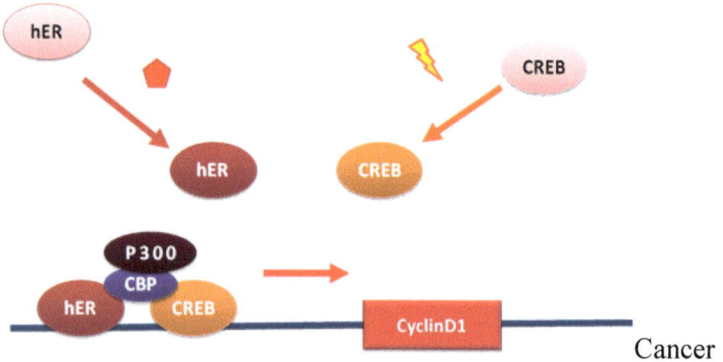

Cancer

As the result, both estrogen and stress will increase the chance to get breast cancer.

Proper Food

Choose Pregnant Age

From : http://www.cancer.gov/cancerinfo/pdq/treatment/breast-cancer-and-pregnancy/healthprofessional/

Breast cancer is the most common cancer in pregnant and postpartum women, occurring in about 1 in 3,000 pregnant women. The average patient is between 32 to 38 years of age and, with many women choosing to delay childbearing, it is likely that the incidence of breast cancer during pregnancy will increase.

Abortion, estrogen and breast cancer

Scientists and researchers have collected data since 1957 that indicates early abortion is a risk factor for breast cancer, yet this information is not well known to the public nor is it acknowledged by many medical associations and journals

Part V
The Future Prevention and Treatment

The Desired Treatment of Breast Cancers

It seems that most breast cancer patients have survived from the current treatments and lived for a very long time without relapse. For a while, I felt rather relaxed about breast cancer. I thought that it is not a fatal disease.

However, I changed mind after my neighbor lady, Emma, died 4 months after she was diagnosed with breast cancer. Emma was a 74 years old single lady. She was kind and funny. She grew a lot of roses at her front door. I often chatted with her when she was watering her roses. Sometimes she invited me to visit her home to talk about many different topics. She took care of herself very carefully. She told me that she never leave food in the room temperature for hours because she didn't want to get cancer. I thought that she was over cautious. One day, she came back from New York, her hometown. She told me that she got cancer. I was shocked. I said that maybe she would be OK because most people could survive from cancers. She said that it was too late. Her cancer that originated from breast had spread all over her body. She showed me that the hard lumps were all over the places under her skin. The doctor told her that she could only live for another 6 months even if she took treatment. She started chemotherapy. She was so weak after the treatment started. Because she couldn't take care of herself anymore, she stayed in a health care center. I went to visited her and brought her a big bunch of roses. She could hardly speak but I could tell that she was happy to see me. A few days later, I heard that she died. I felt so sad. Every time I saw roses at her front door, I felt guilty because I was studying breast cancer and supposed to be able to find some cure for her.

I knew that cancers are still killing people. Emma was the only person died under my watch. That was why it had a bigger impact on me.

Cancers are still killing people. I keep telling myself.

Apparently, the current treatment could not save everyone's life. How should we treat cancer in the future? The real question is how should we successfully deal with cancer without pain, and still live a healthy live?

I asked myself: what I want to do if I got cancer.

I started to think about the treatment for myself before I get the disease.

Here is what I am thinking:

First all, I need understand the disease better. What is cancer? As we discussed before, cancer cells are the cells that have lost ability to stop dividing. They are not necessary to grow fast. They are just not dying.

Why do people think cancer cells are dangerous? If these cells are keeping dividing, we might just have a big breasts or a big piece of lump.

We were dreaming! Cancer cells would not help us to grow big desirable breasts. Instead, they would grow into pieces of lump. After a while, the cells from the lumps would spread all over our bodies: brain, liver, kidney, bone and lung. The organs would lose functions. We would die.

Wait a second, I was pondering. Are you saying that the cancers won't cause death directly, it was the organ failure caused death?

Yes, you are right. That was what happening.

OK now, here is what I want to do:

1. Prevent cancer to occur before it happens;

2. Control it when it is still small and not yet spread;
3. When the caner spread, stop it by using strong medicine.

    You must think all of these are cliché: we know it! You are right. However, here I will mention how in a detail.

1. How to prevent breast cancer to occur before it happens:
    a. Analyze yourself
        i. Did your family have many members have cancers?

            May be your father's family has cancer history, not your mother's, or vice verse.

        ii. If there is any family cancer history, what types of cancer they had? Did the cancer progress quickly or took a long time to progress?

        iii. Do you have a cancer history?
        iv. Do you have a big breast?

            Big breasts are very desirable for women. They are beautiful! Many women spend a lot of time and money to add size on their breasts in order to make them look more attractive. However, at the other hand, big breasts indicate high levels of hormone activity or high level of estrogen receptor concentration in breast cells.

            If you have big breasts, very likely, you have either high level of estrogen or estrogen receptor.

        v. If you have menopause, did it came later than 52 years old of your age?
        vi. Do you have long history overweight?
        vii. Do you allergic to anything?
        viii. Are you tall?

ix. What is your race?

x. Do you take mammography test every year?

In 1998, I visited a hospital to have a physical examination. One of the required procedures was to take a Mammography exam for breast cancer screen. I wasn't happy about it because I didn't want to receive more X-rays. The doctor convinced me that it is a necessary step to detect breast cancer. I was studying breast cancer at that time. I was very scared about the high risk of breast cancer occurrence in US: one out of seven women has breast cancer. 'OK. I will do it." I thought.

I went to the Mammography room. While the doctor assistant filling up the information, I had a brief chat with her.

"It seems that there are so many women get breast cancer in US." I said.

"Yes. Wasn't like this before we had Mammography. It was only about one out five before. After we had this technology, it seems the ratio of breast cancer occurrence increased so much. It is one out of seven now. We think it might be because we can discover more patients now." She said.

I was wondered. What she said is coincident with my concerns.

There was a small area on the Mammography film looks suspicious. The doctor couldn't make conclusion based on the first film. I have to take the second one. When I was sitting in the waiting room, six other ladies were also sitting in the room waiting the calls. They were chatting.

"I found my cancer in a shower. I did Mammo right before that and didn't find anything." A lady said.

"I find mine by myself too. I heard that about 80% of breast cancers are found by self exam." Another lady agreed.

There were many studies on Mammo effectiveness. The results were not consistent. Some study showed that death rate of breast cancer was decreasing since we had the Mammo, while others showed that the total death didn't change. I guess that it is true that Mammo could discover some cancers in earlier stage so that the treatment could be more effective. I haven't seen any study showing whether the Mammo could induce cancer.

I want to discuss with you about whether it is beneficial or harmful to use Mammo.

In the following graph,

    a. No radiation and P53 is not active and cells are growing.

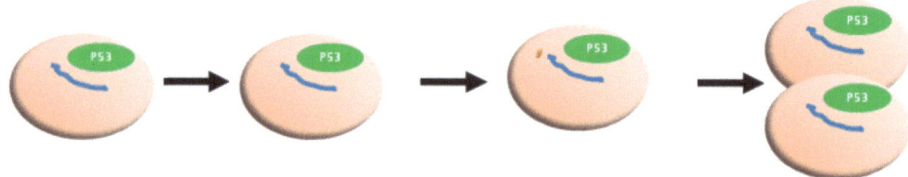

    b. Apply radiation to damaged DNA and P53 will kill the cells:

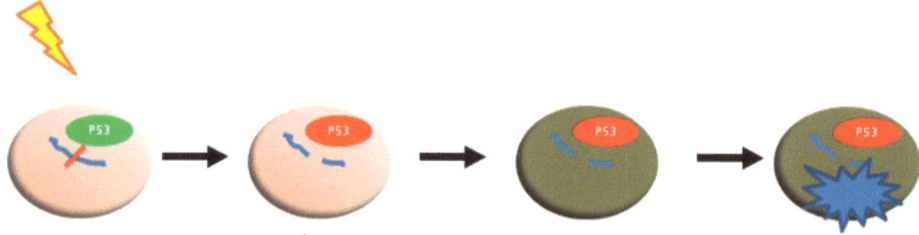

    c. Apply radiation and damaged both DNA and P53, P53 can't stop cells from growing

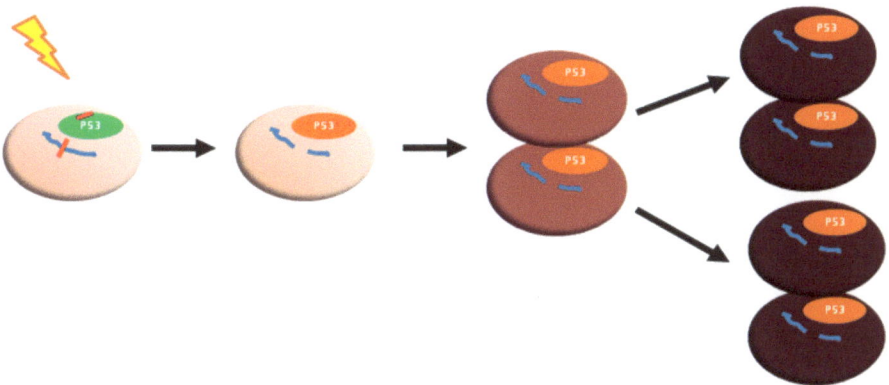

It doesn't sound too bad. However, it is a little misleading If we use the dose instead of using density. Let's look at the density. If we calculate what the density of the radiation is, we can see the difference. Let's assume the time length of the radiation is 10 min even if usually is shorter than that. Three months equals 3 x 30 days x 24 hours x 60 minutes = 129, 600 minutes!! This means that the radiation density is at least 12,960 times higher than the background radiation density.

I did realize this huge difference between Mammo and the background radiation. In the future, I would think about it 3 times before I take Mammograph.

The following is the summary of Mammography exam:

1. Mammo is low dose X-ray but it is high density;

2. Mammo is targeting breast tissues that is very vulnerable to be cancerous;

3. In case you have low DNA repair ability, Mammo will likely cause cell proliferation;

4. In any case, X-ray is a source for DNA damage and thus for cancer.

The non-tissue specific cell dividing rate is faster than the death rate. Growth is faster than the die.

From the previous discussion, we have understood that the cells in breast tissues are constantly growing and dying. The difference between the normal cell and an early stage cancer cell is not that the growing rate of the cell is faster than the death rate.

Instead, it is that the cancer cells have ability to divide unlimited times while the normal cells can only divide about 50 times. In later stage of cancer cells, the growing rate surpasses the ability of differentiation, so that the cells become embryonic.

P53 is playing a commanding role in inducing apoptosis. Mitochondria are believed to be the central stage in p53-mediated apoptosis.

Disclaimer: Anything mentioned in this book is the opinion of the author. It is not medical advice. It is your own responsibility to make decision for your health care issues.

God made Eve with Love. HE wanted us to be healthy. We need to find out how HE arranged for our health. -- Xiaohong

www.ingramcontent.com/pod-product-compliance
Lightning Source LLC
Chambersburg PA
CBHW051057180526
45172CB00002B/674